Real-Life Math

DATA ANALYSIS

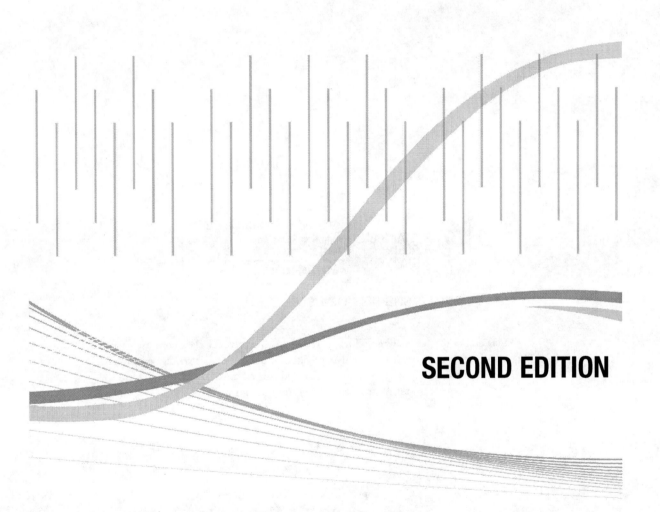

SECOND EDITION

WALCH  PUBLISHING

Certified Chain of Custody
Promoting Sustainable
Forest Management

SUSTAINABLE FORESTRY INITIATIVE

www.sfiprogram.org

SGS-SFI/COC-US09/5501

1   2   3   4   5   6   7   8   9   10

ISBN 978-0-8251-6322-7

Copyright © 2000, 2007

J. Weston Walch, Publisher

P. O. Box 658 • Portland, Maine 04104-0658

www.walch.com

Printed in the United States of America

WALCH PUBLISHING

Table of Contents

Introducing Data Analysis

Data Collection

Basic Data Analysis

Advanced Data Analysis

How to Use This Series

The *Real-Life Math* series is a collection of activities designed to put math into the context of real-world settings. This series contains math appropriate for pre-algebra students all the way up to pre-calculus students. Problems can be used as reminders of old skills in new contexts, as an opportunity to show how a particular skill is used, or as an enrichment activity for stronger students. Because this is a collection of reproducible activities, you may make as many copies of each activity as you wish.

Please be aware that this collection does not and cannot replace teacher supervision. Although formulas are often given on the student page, this does not replace teacher instruction on the subjects to be covered. Teaching notes include extension suggestions, some of which may involve the use of outside experts. If it is not possible to get these presenters to come to your classroom, it may be desirable to have individual students contact them.

We have found a significant number of real-world settings for this collection, but it is not a complete list. Let your imagination go, and use your own experience or the experience of your students to create similar opportunities for contextual study.

Foreword

Every person who pays any attention to the media is bombarded by data. Reports about the effectiveness of new medical treatments, financial market results, hundreds of sports scores, and many other numerical items flood newspapers, airwaves, and the Internet. Included in these reports are key numbers that measure, summarize, and communicate essential facts about the data. These key numbers are *statistics*.

Many people can rattle off lots of detailed statistics about sports, cars, or other areas of high interest. But trying to understand the mathematics underlying statistics puts most people to sleep. This book attempts to deepen students' mathematical understanding of statistics by presenting supplemental activities focused on subject matter with natural appeal—real-life examples and real data about the students themselves. It will fit into a general math, pre-algebra, or algebra course.

After mastering the activities in this updated second edition of the book, students will have many new tools to help them wade through the voluminous statistics they will encounter. In the process, they will learn how to make better decisions in life.

—Eric T. Olson

1. Statistics: How to Measure Anything

Context
daily life

Math Topic
concept of statistic

Overview
There are innumerable ways that statistics are used in everyday life. In this activity, students are asked to make observations about events that have a statistical component.

Objectives
Students will be able to:

- define *statistic*

- identify things in everyday life that involve statistics

Materials
- one copy of the Activity 1 handout for each student

Teaching Notes
To get students started thinking about statistics, ask them to come up with a list of data sets from media reports, sports information, or personal data from everyday life that involve statistics.

There are many, many phenomena that can provide statistics. In fact, nearly any observation of daily activities can be measured with statistics.

Begin with a discussion of the concept of a statistic. Statistics are quantities that convey measurements about groups of data.

These might include a total or sum, an average or mean, a rate, a median, a range, a mode, a variance, or a standard deviation.

Answers

1. This table is meant to have short entries. You might expand on the individual items in a class discussion. At this point, students will probably give examples that involve sums, averages, and rates. Additional examples will vary.

Feature	Statistics involved (including related factors)
1. climate	average temperature, average precipitation
2. car accidents	speed, rate per vehicle-mile traveled, average blood-alcohol level
3. class grades	average test scores, spread of test scores (curve)
4. customers at mall	total sales, sales per customer, profit
5. state lottery	total sales, total jackpot

2. A statistic is any number that conveys a measurement about a set of data. For example, a store may record daily sales by adding together the amount of every sale during an entire day. The sum is a statistic. Likewise, a store may add up sales for every day during a year and divide by the number of days to get the average sales per day. The average is a statistic.

Extension Activity
Have students collect examples of statistics used in real life and report back to the class. The Internet, newspapers, and newsmagazines are great sources.

1. Statistics: How to Measure Anything

1. The table below lists five features of daily life that have statistical aspects. Explain what statistics are involved. Then write five more examples of your own.

Feature	Statistics involved (including related factors)
1. climate	
2. car accidents	
3. class grades	
4. customers at mall	
5. state lottery	
6.	
7.	
8.	
9.	
10.	

2. Explain what is meant by the term *statistic*. Give an example.

2. Gathering Data: Reaction Times

Context

driving

Math Topic

data collection

Overview

There is no better way to understand statistics than to do simple experiments. This activity allows students to generate a small data set by measuring their own reaction times. The data collected here will also be used in later activities. They are a prerequisite for Activities 8, 12, and 13.

Objectives

Students will be able to:

- collect a set of data in a simple experiment

- describe experiences during an experiment in qualitative terms

Materials

- one copy of the Activity 2 handout for each student

- one 12-inch (30-cm) ruler for each pair of students

- calculators

Teaching Notes

The idea in this activity is to collect some real experimental data that students later may analyze statistically. It is put in the context of research concerning driving.

The experiment involves two students. One student sits, while the other student stands behind the sitting student. The standing student holds a ruler vertically with thumb and forefinger near the 12-inch end so that the 1-inch end points downward and is visible to the sitting student. The students should be positioned so that the sitting student can see the 1-inch end of the ruler but cannot see the standing student's hand. The sitting student should then place whichever hand he or she wants in a grasping position at the 1-inch mark near the bottom of the ruler, but not touching the ruler. The sitting student should use the upper part of his or her horizontally positioned thumb to align with the 1-inch mark of the ruler.

To conduct a trial, the standing student should carefully drop the ruler at a random moment. The sitting student should then catch it as quickly as possible. The sitting student should record the distance the ruler dropped before he or she caught it. Note that this distance is equal to the change of position from the initial alignment to where the ruler was caught. If students initially align their thumbs with the 1-inch mark,

(continued)

2. Gathering Data: Reaction Times

the change of position will be equal to the final position less 1. This distance will be converted into the time required for the sitting student to react to the falling ruler. Students should conduct 15 trials, switch roles, and conduct 15 more trials. They should record their own results on their own copies of the handout.

After all the distance data are collected, students should convert them to time data. The falling ruler can be considered to be an object in free fall without air friction. Thus, if d represents the change of position in inches, then t (time in seconds) can be calculated using this formula:

$$t = (0.072) \sqrt{d}$$

If d is in centimeters, then t may be calculated as follows:

$$t = (0.045) \sqrt{d}$$

Answers

1. Results will vary considerably. Make absolutely sure that all students record change of position, not merely the final position. Help students calculate the correct values for the actual reaction time using the formulas above.

2. Not all trials will proceed perfectly. Two common occurrences will be premature anticipation and lack of attention by the subject (sitting student). Students should record any such events. These notes will become important when students are asked to explain and deal with outliers in the data.

Extension Activities

*   Have students write a paragraph explaining how the Department of Motor Vehicles might use data about people's reaction times.

*   Have students write a procedure for an experiment to test people's reaction times under different conditions. For example, how would reaction times be affected by dim light, or by a bright light shining in the eyes? What about sudden loud noises? If you wish, have students perform some of these experiments to see how their reaction times change, and whether their predictions of the effects of different conditions were correct.

2. Gathering Data: Reaction Times

Imagine you work for the State Department of Motor Vehicles and are studying people's reaction times. (The ability to react quickly is very important while driving.) You want to test a simple experiment requiring very little equipment that you think may give you good results.

1. Work in pairs. One partner will be seated. The second partner will stand behind the seated one while holding a ruler in front of the sitting partner. Carefully follow your teacher's instructions about how to conduct a trial. At a random moment, the standing partner will release the ruler. The seated partner will catch it. You will measure the ruler's change of position between the moment of release and the moment the ruler is caught. The sitting partner should not be able to see the standing partner. Careful measurements are required! Make sure the sitting partner correctly aligns and records thumb position before and after the trial. Do 15 trials, switch roles, and do 15 more. The sitting partner should record his or her own data in the Change of Position column on his or her own handout. Your teacher will explain how to calculate the value for reaction time. Time is measured in seconds.

    The change in position shows, in inches, how long it took you to react and catch the falling ruler. Now you need to convert this distance measurement to a time measurement. To find the time in seconds, use the following formula:

    $$t = (0.072) \sqrt{d}$$

    Example: If your thumb was at the 2.25-inch mark when you caught the ruler, then d (the distance the ruler traveled) is 1.25 inches. To calculate the time it took you to react, find the square root of d: 1.118. Now multiply 0.072 by 1.118. The answer is 0.08 seconds.

(continued)

2. Gathering Data: Reaction Times

Trial	Change of position	Reaction time $t = (0.072)\sqrt{d}$	Trial	Change of position	Reaction time $t = (0.072)\sqrt{d}$
1			9		
2			10		
3			11		
4			12		
5			13		
6			14		
7			15		
8					

2. Explain any problems you had during your trials. Did the partner being tested (the sitting partner) ever improperly anticipate or forget to react to the ruler falling? How did you solve these problems? Which trials were affected?

_____

_____

_____

_____

_____

_____

_____

_____

_____

_____

_____

_____

_____

3. Radio Preferences

Context

radio

Math Topic

sampling

Overview

We live in an era where our reactions to products and to media are heavily studied. The results are then used to determine what products or media will be produced.

In this activity, students compile and analyze information about music preferences to help a new radio station decide what type of music to play during their radio shows. During this process, they learn how statistics are gathered with a survey reflecting a sample of a population.

Objectives

Students will be able to:

- compile and interpret data from a survey

Materials

- copies of the four Activity 3 handouts for each student

Teaching Notes

Ask students why the radio station would be interested in learning what music they listen to. Lead students to the realization that commercial radio stations play music to get a good audience; once they have that audience, they can get advertisers. You can also help them understand that advertising dollars support the radio station and allow the audience to listen to the music for free.

Discuss with students how surveys are used to collect data in real life. Some students may have seen data collectors standing in a mall with a clipboard, asking people questions as they walk by. Some students might also have participated in online surveys.

Emphasize that surveys are different from questionnaires. In questionnaires, people are generally given a limited number of choices for their responses. In surveys, people are asked a question and allowed to respond in any way they choose.

At this point, students may be able to understand why a survey works. In this instance, the radio station is trying to learn what young adults like to listen to and then break that information down for each of two types of advertisers.

Once individual students have completed the data gathering part of the activity, have them share their responses with the rest of the class. One approach is to prepare an

(continued)

teacher's page

3. Radio Preferences

overhead transparency of the Consolidating Data sheet. Appoint one student to record the data on the overhead, using tally marks. Then have each student name his or her three favorite songs and buying preferences. The recorder should write the name of the song in the left-hand column, then put a tally mark in the appropriate Buying Preference column. If the same song (as is likely) is named again, the recorder just makes another tally mark in the appropriate column. See the example below.

Song	Video games	Clothes	Total
"Crazy"	| |	| |	| | | |
"Satellite"	| | | |	| | | |	~~||||~~ | | |

All students should copy the data onto their own Consolidating Data sheets. Alternatively, you may have students work in small groups to begin the data consolidation, then share the group data as described above.

Once all students have completed the Consolidating Data sheets, they can work alone or in small groups to analyze the data.

Answers

1–6. Answers will vary.

7. Students should conclude that the survey was a better choice because it allowed respondents to choose their favorite songs rather than pick from a list of songs they may not like all that much.

Extension Activities

- Ask students what the radio station should do if it decides to try to get automobile dealers to advertise. Would a new survey be necessary, or could the station use data it already has? Would it be advisable to target older people who can afford cars? Have students write up a plan of action for the advertising manager.
- Choose the three most popular songs on the list. Have each student choose a favorite song from among those three, then create a pie chart of class choices based on the result, either manually or using a spreadsheet program.

3. Radio Preferences

Gathering Data

A new radio station is about to go on the air. The advertising director wants to target young adults as their audience. The music director must come up with a playlist that meets this demand. They have asked you to complete a survey, listing your three favorite songs and an indication of your buying preferences.

Music Survey

Which item are you more likely to buy?

☐ Video games   ☐ Clothes

Please list your three favorite songs below, in order of preference:

1. _____

2. _____

3. _____

(continued)

3. Radio Preferences

Consolidating Data

Use the form below to record the entries for the survey.

   The first time each song is named, write it in the left-hand column. For each song, put a tally mark in the appropriate Buying Preference column. If the same song is named by another survey respondent, do not write the name again. Just put a tally mark in the Buying Preference column next to the song title.

   When you have listed all the songs named by respondents, total the responses.

   a.   Total all the responses in the Video Games column. Write the total at the bottom of the column.

   b.   Total all the responses in the Clothes column. Write the total at the bottom of the column.

   c.   Find the total number of respondents who chose each song by adding the tally marks in both the Video Games and Clothes columns on that row. Write the total in the right-hand column.

Song	BUYING PREFERENCES		
	Video games	Clothes	Total
Total			

(continued)

3. Radio Preferences

Analyzing Data

Use your findings to answer the following questions.

1. Which song seems to be the most popular? Is this result significant?

2. If the advertising director decides to target video-game buyers, which five songs should the station include on its playlist?

3. If the advertising director can get only clothing stores to advertise on the station, which five songs should be included on the playlist?

4. Look at the most popular song on the survey. Was that song preferred by more video-game buyers or more clothes buyers? Is this information significant?

5. If you were the music director, what songs would you include on a playlist to target clothes shoppers on Saturday afternoons?

6. Plan a playlist that targets video-game buyers on Sunday afternoons.

7. Why did the advertising director choose to use a survey rather than a questionnaire?

4. Television Preferences

Context

television

Math Topic

sampling

Overview

In this three-part activity, students keep a diary of their television viewing habits for one week. They then compile this information to help a local college decide which television shows would yield the best response to its advertisements. During this process, they learn how statistics are gathered with a representative sample.

Objectives

Students will be able to:

- maintain a diary of their viewing habits

- compile and interpret the data from all students' answers

Materials

- copies of the three Activity 4 handouts for each student

Teaching Notes

Ask students why the local college would be interested in learning which programs they watch. Lead students to the realization that they are the audience these commercials want to reach. You can also help them understand that advertising dollars are precious, and the college does not want to waste money on commercials during shows that high school students don't like.

Once individual students have completed the data-gathering part of the activity, have them consolidate their data. One approach is to prepare an overheard transparency of the Consolidating Data sheet. Appoint one student to record the data on the overhead, using tally marks. Then have each student name the TV shows he or she watched. (The handout has space for three shows in each time slot. You may need to adjust this, depending on how many different shows students watch in each time slot.) The recorder should write the name of each show in the TV show column, then make a tally mark in the Viewers column for each student who watched that show.

Once all students have completed the Consolidating Data sheet, they can work alone or in small groups to analyze the data.

Discuss with students how diaries are used to collect data in real life. Many students have probably heard of television ratings, sweeps month, or Nielsen ratings. Some students may even be able to name some of the top-rated shows on television now.

Emphasize that the Nielsen ratings are gathered using a representative sample.

(continued)

4. Television Preferences

Nielsen Media Research uses a sample of approximately 25,000 households to represent the television viewing habits of the 111 million households with television watchers. Students may speculate that 25,000 households cannot possibly be a large enough sample to work. But Nielsen Media chooses these households carefully, and their ratings have been accurate enough to keep them in business for over 50 years.

At this point, students may be able to understand why a representative sample works. For national viewing habits, set meters are attached to television sets. Nielsen Media processes approximately 10 million viewing minutes a day. During sweeps periods, they collect and process additional data from 1.6 million paper diaries from households across the United States.

You may choose to have your students keep their diaries for a full week, or you may have them fill out their diaries from memory during the class period.

Answers

All answers will vary.

Extension Activities

- Have students survey other students in the school to see if their favorite television shows are similar to those of this class.

- Have students go to the Nielsen Media Research web site at www.nielsenmedia.com to find out more about how television ratings work. The explanation of representative sampling is well presented.

- Have students use the Internet to find daily television ratings for the period they are analyzing. How do the class results compare to national results? Have students create a graph, either manually or using a spreadsheet, to compare both sets of results, the national ratings and classroom ratings.

4. Television Preferences

Gathering Data

A local college needs to increase enrollment. To pique interest in the school, college administrators have decided to advertise during popular television programs. They can afford to advertise only during the first hour of prime-time television programming, which is from 8:00 P.M. to 9:00 P.M. eastern standard time and Pacific daylight time, or 7:00 P.M. to 8:00 P.M. central daylight time and mountain standard time.

The college first needs to find out which shows are popular with local high school students. In this activity, you will keep a diary of five days' worth of television viewing. Use the diary provided below. If you do not watch television during any given time slot, simply leave that line blank. All times below are given in eastern standard time. If you live in another time zone, use the equivalent time slot for your zone. For each time slot, you can enter only one program.

Day	Time	TV show
Monday	8:00–8:30 P.M.	
	8:30–9:00 P.M.	
Tuesday	8:00–8:30 P.M.	
	8:30–9:00 P.M.	
Wednesday	8:00–8:30 P.M.	
	8:30–9:00 P.M.	
Thursday	8:00–8:30 P.M.	
	8:30–9:00 P.M.	
Friday	8:00–8:30 P.M.	
	8:30–9:00 P.M.	

(continued)

4. Television Preferences

Consolidating Data

Use the form below to record all entries for the diaries kept in the previous part of this activity. While you will collect the responses as a class, you should each record the responses and totals on your own copy. (If you are in a time zone other than eastern standard time, you may have to adjust the time slots on this sheet.)

Day	Time	TV show	Viewers	% of Total Viewers
Monday	8:00–8:30 P.M.			
	8:30–9:00 P.M.			
Tuesday	8:00–8:30 P.M.			
	8:30–9:00 P.M.			
Wednesday	8:00–8:30 P.M.			
	8:30–9:00 P.M.			
Thursday	8:00–8:30 P.M.			
	8:30–9:00 P.M.			
Friday	8:00–8:30 P.M.			
	8:30–9:00 P.M.			

(continued)

4. Television Preferences

Analyzing Data

Use your findings to answer the following questions.

1. Now that you have collected all the responses on your record sheet, present the data in a useful manner. Count up all the responses for each TV show and put the total in the appropriate column.

2. a. Which show had the largest number of viewers?

   b. Which show had the smallest number of viewers?

3. Add up the number of students who watched television on Thursday from 8:00 to 8:30 P.M.

   a. How many students watched television in that time slot?

   b. What was the most popular show in that time slot?

   c. What percentage of all student viewers watched that show?

4. Randomly gather about one-fourth of all diaries in your class. Add up the number of students in that representative sample who watched television on Thursday from 8:00 to 8:30 P.M.

   a. How many representative students watched television in that time slot?

   b. What was the most popular show in that time slot?

   c. What percentage of all representative students watched that show?

   d. Is the answer to 4c similar to the figure found in 3c?

(continued)

4. Television Preferences

5. If the college could afford to advertise during one show only, which show should it be? Explain your answer.

6. Assume that the college cannot afford to advertise during that show. Which show should be the college's second choice? Explain your answer.

7. Is this the same show that would be chosen by the representative sample?

8. Assume that the college can afford to put one commercial on each of the major networks all on the same night. The college's marketers want to know which night is the most popular night for young adults to watch television. Figure out which night this would be.

5. Movie Preferences

Context

recreation

Math Topic

statistical significance

Overview

In this two-part activity, students collect data for a movie producer who is unsure about how to end the movie. They then compile the data for the whole class and conclude which ending the producer should use for the film.

Objectives

Students will be able to:

- gather data

- compile and interpret data

Materials

- copies of the three Activity 5 handouts for each student

Teaching Notes

Before starting this activity, students should be familiar with the concept of statistical significance.

To get students started thinking about data collection, ask them why a movie producer might want to test different endings before releasing a film. Lead them to discuss how movie endings might vary depending upon the age and gender of the viewers the producer is trying to reach.

Begin with a discussion of the use of observation in collecting data. In some cases, two movie endings are filmed and shown to viewers, and viewers are then allowed to pick their favorite ending. Producers then use these observations to determine which ending would result in the most movie tickets sold.

Emphasize that accuracy is important when collecting this type of data. Producers need to know more than just the overall favorite ending. They also need to know the age group and gender of all reporting viewers in order to determine which ending will please their target audience.

Once individual students have completed the data-gathering part of the activity, have them share their responses with the rest of the class. One approach is to prepare an overhead transparency of the Analyzing Data sheet. Appoint one student to record the data on the overhead, using tally marks. Then have each student give his or her responses to the questions on the survey, while the recorder puts a tally mark in the appropriate column.

Alternatively, you can have students work in small groups to begin the data consolidation, then share the group data as described above.

(continued)

5. Movie Preferences

All students should copy the data onto their own Analyzing Data sheets.

At this point, students may be able to speculate what types of movies are targeted to their age group. These can range from movies with a lot of action and special effects to movies that are targeted as the "perfect date" movie.

Answers

Answers will vary depending on students' preferences.

Extension Activity

Have students discuss movies they have seen in the past few months. Allow them to debate whether or not they enjoyed the ending of the movie or if they, as producers, would have changed it.

5. Movie Preferences

Gathering Data

A producer is about to release a new movie, which she hopes will be a summer blockbuster. The problem is that the director filmed two different endings, and the producer can't decide which one to use.

She has already decided that the target audience for the movie will be young adults. So she went looking for a group of young adults to test the two endings on—and came up with your class. Each of you has been given a summary of the movie plot, with both endings. Read the summary and endings. Then answer the questions that follow.

Down the Mountain

Summary

Adira, her brother Matt, and Matt's friend Josh are spending a long weekend hiking and climbing in the Rockies. All three are experienced climbers and have spent a lot of time in the wilderness. But this is the first time the three of them have done any climbing together, and by the second day out, tensions have started to build. Josh is strongly attracted to Adira and tries to get her attention by taking more risks than he usually takes when climbing. Matt is irritated that Josh is showing off and is jealous of the attention Josh is paying Adira. Adira, who is not familiar with Josh's usual climbing style, is aware that something is causing tension, but isn't quite sure of the source.

By the third day, Matt is barely speaking to either Josh or Adira. Adira, hurt and confused, is happy about the attention Josh gives her. This makes Matt even angrier. As they approach their last ascent of the day, Matt is leading the climb, with Adira second and Josh bringing up the rear. In his anger, Matt is climbing carelessly.

Ending 1

Matt accidentally dislodges a rock, which bounces off the mountain and hits Adira. Adira slams backward into the rock wall, then swings out again and dangles helplessly from the rope, her right leg clearly broken. Using the ropes, Matt is able to lower Adira to Josh, but Adira cannot climb back down the mountain. Matt sets and splints her leg. Then he and Josh create a makeshift stretcher

and begin the arduous task of carrying Adira down the ascents they had all toiled up earlier. The final scene shows Adira in the hospital while Josh and Matt, their friendship restored, leave the building together.

(continued)

5. Movie Preferences

Ending 2

Two-thirds of the way up the rock wall, Matt misjudges a foothold and tumbles from the cliff. The ropes holding the three climbers together break his fall, but he is dangling, unconscious, from the rope. Josh and Adira together are able to haul him to the relative safety of a rock ledge, but they know they cannot carry him down the mountain alone. They spend the night sitting on the ledge, not daring to sleep in case they fall. Instead, they talk and grow closer through the night. At first light, Josh sets off to get help, leaving Adira with Matt. At last, Matt regains consciousness. He is in pain, but knows Adira, and admits that his own recklessness caused the accident. Josh arrives with a rescue team, and they begin to carry Matt down from the mountain. Josh and Adira, now linked closely together, follow.

Question	Response	
1. Which ending did you prefer?	☐ Ending 1	☐ Ending 2
2. Is this the type of movie you would go see in a theater?	☐ yes	☐ no
3. What is your age group?	☐ under 15	☐ 15 or older
4. How many movies, on average, do you see each year?	☐ 12 or more	☐ 11 or fewer
5. What is your gender?	☐ male	☐ female

(continued)

21

5. Movie Preferences

Analyzing Data

Use the chart below to record the responses from the movie questionnaire.

Possible choices	Male responses	Female responses	Totals		
			M	F	All
1. Ending 1					
Ending 2					
2. yes					
no					
3. under 15					
15 or older					
4. 11 or fewer					
12 or more					

1.   Now that you have collected all the responses on your record sheet, present the data to the producer in a useful manner.

   a. Count all the male responses to each choice and enter the totals under the *M* column.

   b. Count all the female responses to each choice and enter the totals under the *F* column.

   c. Count all the responses for each choice regardless of gender. Put these totals under the *All* column.

2.   Which ending did more of the class prefer? Is this result significant?

3.   If the movie producer decides to target teenage girls with this movie, which ending should she use for the film? Is this result significant?

4.   Did more males or more females indicate that they would see this type of movie in a theater? Is this result significant?

5.   Is the movie producer more likely to use your data if most of the students in your class say they see more than a dozen movies per year? Why or why not?

6. Food Preferences

Context

food

Math Topic

sampling

Overview

In this two-part activity, students begin by collecting data for neighbors who are about to open their own pizzeria. They then compile these data for the whole class and turn them into useful information for the restaurant owners.

Objectives

Students will be able to:

- answer a questionnaire
- compile and interpret data

Materials

- copies of the three Activity 6 handouts for each student

Teaching Notes

To get students started thinking about data collection and market research, ask them what a new restaurant might need to know before opening in your community.

Begin with a discussion of the concept of data collection. If asked to name what a restaurant may need to know before

compiling a menu, students should be able to come up with some basic questions, such as what kinds of beverages to serve, and whether or not soups and salads are popular in the area.

Emphasize that data can be collected in a variety of ways. If the restaurant owners know what questions to ask, they can get local residents to answer questionnaires. Ask students if they are familiar with the term *data processing*. This is what happens when businesses compile information into a format useful to them. This process is usually done on a computer.

Once individual students have completed the data-gathering part of the activity, have them share their responses with the rest of the class. One approach is to prepare an overhead transparency of the Analyzing Data sheet. Appoint one student to record the data on the overhead, using tally marks. Then have each student give his or her responses to the questions on the questionnaire, while the recorder puts a tally mark in the appropriate column.

Alternatively, you may have students work in small groups to begin the data consolidation, then share the group data as described above.

All students should copy the data onto their own Analyzing Data sheets.

At this point, students may be able to speculate what a business would do with its

(continued)

6. Food Preferences

data once they are compiled. Students may be familiar with graphs and pie charts from other mathematics classes. Emphasize that these types of charts are used often in businesses, and for more than simply planning menus.

Answers

1. The term *questionnaire* describes a list of questions planned beforehand and used to collect data. In some instances, it is best to limit the number of possible answers on a questionnaire to keep the collected data more manageable. Imagine if every student had a different favorite pizza topping—the restaurant's menu would have to be ten pages long!

2. This table is designed so that students can pick from the given choices. You might expand upon the individual items in a class discussion.

Answers to the data analysis portion of the activity will vary.

Extension Activities

*   Have students create a menu for the pizzeria. Ask them to come up with five points they would emphasize if they were advertising this restaurant in the local newspaper.

*   Have students create a spreadsheet using the questionnaire data and generate graphical representations of the data to present to the restaurant owners.

6. Food Preferences

Gathering Data

A family in your neighborhood is known for creating some of the best pizza in town. Because of this, they have decided to buy a building and open a pizzeria. You suggest using a questionnaire to help them plan their menu.

You say, "A questionnaire would be very helpful."

1. First, you need to explain to your neighbors just what a questionnaire is, and why it is important to limit the number of choices you can select on the questionnaire. Write your explanation below.

2. Your neighbors ask you and your classmates to help them. Here is the questionnaire they develop. Answer the questions by giving your personal preferences.

Question	Your response	
1. What are your two favorite pizza toppings?	☐ pepperoni ☐ mushrooms ☐ sausage	☐ extra cheese ☐ ham and pineapple ☐ other _____
2. Do you prefer a thin crust or a thick crust? Please choose only one.	☐ thin crust ☐ no preference	☐ thick crust
3. Do you prefer stuffed crust or plain?	☐ stuffed crust ☐ no preference	☐ plain crust
4. Do you like your pizza cut into slices or squares? Please choose only one.	☐ slices ☐ no preference	☐ squares
5. What kind of beverage do you like best with pizza? Please choose only one.	☐ soda ☐ water ☐ milk	☐ diet soda ☐ fruit juice ☐ other

(continued)

6. Food Preferences

Analyzing Data

Use tally marks to record the responses from the questionnaire on the chart below. While you will collect the responses as a class, you should each record the responses and totals on your own copy. Then answer the questions on the next page.

Possible choices	Student responses	Totals	Percent of total
1. pepperoni			
extra cheese			
mushrooms			
ham and pineapple			
sausage			
other			
2. thin crust			
thick crust			
no preference			
3. stuffed crust			
plain crust			
no preference			
4. slices			
squares			
no preference			
5. soda			
diet soda			
water			
fruit juice			
milk			
other			

(continued)

Real-Life Math: Data Analysis

6. Food Preferences

1. Now that you have collected all the responses on your record sheet, you need to present the data in a useful manner. Count all the responses for each choice and put the total in the appropriate column.

2. a. What were the two most popular choices for pizza toppings?

   b. Should the pizzeria offer these pizza toppings on their menu?

3. Did your class prefer thick-crust or thin-crust pizza? Judging by the data you collected, should the pizzeria offer thin-crust pizza, thick-crust pizza, or both?

4. Did your class prefer stuffed crust or plain? Judging by the data you collected, determine which crust the pizzeria should offer.

5. Did your class prefer slices or squares? Judging by the data you collected, should the pizzeria decide how to cut the pizza based on the preference of your class or on the method easiest for the restaurant staff?

6. What beverages should the pizzeria offer?

7. Music Preferences

Context

music

Math Topic

sampling

Overview

In this two-part activity, students learn how to compile data from responses to two different questions into one Top Five Chart.

Objectives

Students will be able to:

- come up with potential sales figures and use them in a meaningful way

- create a chart based on the figures they come up with

Materials

- copies of the three Activity 7 handouts for each student

Teaching Notes

Tell students that in this activity, they will gather data from two sources and compile it into one chart. In this case, they will be creating a Top Five Chart for the most popular singers and groups.

Ask students why gathering data from more than one source makes sense. Begin the discussion by explaining that real music charts are created using data gathered from random national samples of sales reports, radio playlists, and monitored radio. Lead students to the conclusion that if the companies who compile these charts used only sales data, they would exclude the people who listen to songs on the radio. Also, if they got their data only from radio playlists, they would be excluding those who purchase songs and albums.

Ask students why it is important to use both groups. Some students may not know that performers make money not just from song and album sales, but also from royalties received every time their song is played on the radio. So, even those who do not buy albums can contribute to the popularity of a group by listening to and requesting songs on the radio. At this point, students should understand why gathering the data from more than one source is important in this case.

Once individual students have completed the data-gathering part of the activity, have them share their responses with the rest of the class. One approach is to prepare an overhead transparency of the Analyzing Data sheet. Appoint one student to record the data on the overhead, using tally marks. Then have each student give his or her responses to the questions on the questionnaire, while the recorder puts a tally mark in the appropriate column.

(continued)

7. Music Preferences

Alternatively, you may have students work in small groups to begin the data consolidation, then share the group data as described above.

All students should copy the data onto their own Analyzing Data sheets.

Answers

All answers will vary depending on students' responses.

Extension Activities

- Point-of-sale scanners are sometimes used to collect music sales data to create top-album charts. This system is used in the United States and Canada by a company called Nielsen SoundScan. Over 14,000 stores, including retail and online stores, transmit their sales data weekly to compile a chart. Once the chart is created, the computer resets everything at zero, and the process begins again for another week. Have students determine what makes this system better than the random sampling. Have them write about their conclusions.

- Have students prepare a survey on MP3 use in their class or school, then analyze the survey results.

7. Music Preferences

Gathering Data

An online music store wants to start displaying a list of the five most popular singers and groups in your area. Since most of its customers are high school students, your class has been asked to help. To do this, please complete the following survey.

Musical Singers and Group Survey

If you were going to download five songs today, which five singers or groups would you be likely to purchase? Enter your answers below, beginning with your favorite performer.

1. _____

2. _____

3. _____

4. _____

5. _____

Think of five songs you have heard recently on the radio and liked. Write the names of the performers who sang them below.

1. _____

2. _____

3. _____

4. _____

5. _____

(continued)

7. Music Preferences

Analyzing Data

Use the chart below to record the entries for the survey. While you will collect the responses as a class, you should each record the responses and totals on your own copy. The first time a performer is named, write the name in the appropriate space. Make a tally mark for each student choosing that artist. Then answer the questions on the next page.

Download Wish List			Radio Songs		
Performer	No. of students	Total	Performer	No. of students	Total

(continued)

Real-Life Math: Data Analysis

7. Music Preferences

1. Now that you have collected all the responses on your record sheet, create a Top Five Chart by compiling the data from the previous page.

    a. Count all the responses for each performer under Download Wish List and enter the total under the Total column.

    b. Count all the responses for each performer under Radio Songs and enter the total under the Total column.

2. Who was the most popular performer under Download Wish List?

3. Who was the most popular performer under Radio Songs?

4. For each performer who appears on both lists, add up the total responses from the two lists. Which performer has the highest total now?

5. Determine which five performers have the highest totals. The totals do not need to appear on both lists. For example, a performer who gets 20 under Download Wish List and 0 under Radio Songs has a higher total than an artist who gets 10 under Download Wish List and 9 under Radio Songs. Write the top five in the chart below.

Top Five Chart
1.
2.
3.
4.
5.

8. Landmarks in Data

Context

reaction time

Math Topic

data analysis

Overview

Once students have some data in front of them, what can they say about it? This activity introduces a few concepts frequently used to depict the properties of a data set—median, mode, range, and outliers. These properties are usually found quickly without a great deal of calculation. Included are exercises involving the reaction-time data students collected in Activity 2.

Objectives

Students will be able to:

- identify landmarks in a data set

- quickly develop rough ideas about the meaning of data using landmarks

Materials

- one copy of the Activity 8 handout for each student

- Reaction-time data from Activity 2; if students have not done Activity 2, give them the data from the Answer section, in random order.

Teaching Notes

Students have been learning how to collect data. Explain to them that the next step is to learn how to organize that data.

Begin a discussion by asking students if anyone knows what a landmark is. They may assume that they do not know, since in this class *landmark* would be a statistics term. Tell them that they may know more than they think. Ask them not to think about statistics for a moment and then say what they think *landmark* means. Students may come up with a variety of answers, including a tree or other object that marks the boundary of a piece of land, a point in an area that is used to determine the location of other points, or a building with historical value.

Ask them what their definitions all have in common. Lead students to the conclusion that all their definitions contain an object that stands out. Explain that the same is true for landmarks in data sets. The objects that stand out in data sets are the median, mode, range, and outliers. These make excellent reference points when analyzing or referring to the data.

(continued)

8. Landmarks in Data

Answers

1. Here is a set of reaction-time measurements in seconds from Activity 2, ordered from lowest to highest:

0.14	0.15	0.15	0.16	0.16	0.17	0.18	0.18	0.18	0.18	0.18	0.18	0.19	0.20	0.23

2. Answers will vary. Outliers are often the values most quickly identified in a data set. In this set, 0.23 is extreme enough to be considered the result of errors in procedure. For example, the person being tested may not have been paying attention when the ruler was dropped during the reaction-time test. Note here that some judgment is required in identifying outliers.

3. Answers will vary. The sample reaction-time data have a clear modal value of 0.18 seconds. Such a statistic can give the researcher confidence that the experiment is well-run with repeatable results.

4. Answers will vary. It may help to review the definition of *median* here. (With an even number of values, recall that the median is equal to the two values in the middle of the rank order divided by 2.) The eighth value of the sample is in the middle of the 15 total values, so the median is 0.18. The median is a valuable statistic because it is not sensitive to outliers.

5. Answers will vary. In the sample set, the minimum is 0.14, the maximum is 0.23, and the range is 0.09.

6. Answers will vary. In the sample set, the landmarks allow you to quickly identify 0.18 as a valid and repeatable measurement of the reaction time.

Extension Activity

Have students use the Internet to look up statistics for their favorite sports team. Ask them to find the landmarks in these data—the median, mode, range, and outliers.

8. Landmarks in Data

When you obtain a set of data, what can you do to quickly describe it? As you do research on drivers' reaction times, you would like to be able to assess the results and make decisions about the validity of your methods. Finding the main statistical landmarks—median, mode, maximum, minimum, and range—and identifying outliers can help your research.

1. Write your 15 time measurements from the results of your reaction-time tests in Activity 2, ordered from lowest to highest:

2. Take a look at your data. When numbers are far above or far below the averages, these numbers are called the outliers. If there are any, what are the outliers in your data? Were there any reasons that these values occurred?

3. In your data, some values appear more frequently. The most frequently appearing value in a set of data is called the **mode**. Does your data have a modal value? If so, what is it?

4. A quick way to find the "middle" of your data is to find the **median** (equal number of data values above and below this value). What is the median of your data set? What are the advantages of using the median to describe data?

5. The difference between the minimum and maximum numbers in your data is called the **range**. What are the minimum, maximum, and range of your data?

6. Sometimes the most valuable statistical information obtained from your data is the quick description and interpretation you make from the data landmarks. How would you use these landmarks to describe your data?

9. Averages: What Does the *Mean* Mean?

Context

test scores

Math Topic

mean

Overview

Perhaps the most common statistic referred to in daily life is the *average.* **Mean** is technically defined as the sum of the measurements divided by the number of measurements. The average of a series of measurements is equivalent to the mean. However, average is sometimes used to refer to quantities that really are not means— baseball batting averages, for example. Mean is less commonly mentioned in real-life contexts.

The mean may or may not be descriptive of a data set. This activity asks students to judge whether or not the mean is a good representation of a set of data.

Objectives

Students will be able to:

- calculate the mean (average) of a group of measurements

- decide if the mean is a good way to represent a data set

Materials

- one copy of the Activity 9 handout for each student

- calculators

Teaching Notes

Begin discussion by having students think of as many reported averages (means) as they can. Many examples from sports are possible. Per game averages are very common—points per game, rebounds per game, and so on. However, the common baseball statistic, the batting average, is not really a mean by the strictest definition. Batting average is really just a ratio of hits to "official at bats."

Ask students what they would do if they wanted to find out the average grade on a test when the individual grades are 100, 98, 75, 73, 72, and 68. Students should conclude that they would simply add up the scores for a total of 486 and then divide that total by 6. Explain to students that this would give them the mean, which is the average score of 81.

This appears to be a good average grade. However, remind students that this is not necessarily the best representative number. After all, only two students got a grade above the mean, while four students got grades below the mean.

(continued)

9. Averages: What Does the *Mean* Mean?

Answers

1. The mean of these test scores is 83.

2. The mean of these test scores is 83.

3. Answers will vary, but students should conclude that the mean in question 1 is more meaningful because the data is spread evenly around it. In the case of the data from question 2, all values except one fall below the mean—the case where a single score raises the class average above all of the other students' scores.

4. The median for the data in question 1 is 84. The median for the data in question 2 is 79.

5. The median is a better statistic for the data in question 2. The median is less influenced by the single score of 100. That score may be considered an outlier.

Extension Activity

Play a game! Give pairs or teams of students a deck of cards. Have each student or team draw three cards and then calculate the average of the three cards. (Aces are worth 11 and face cards are worth 10.) The next student or team draws three cards and performs the same calculation. The student or team with the highest mean keeps all six cards. The student or team with the most cards when the deck is gone wins.

9. Averages: What Does the *Mean* Mean?

There always seems to be someone in a class of students who raises the class average. So is the average score for a test a good statistic?

The **mean** is a useful statistic. It communicates in a single number the result or effect of a set of data. It is commonly referred to as the **average**. The mean is equal to the sum of all the data values divided by the number of values. Ideally, this number is the mathematical center of the data. But different sets of data can result in means with different meanings, as in the groups of test scores below.

1. Calculate the mean for the following test scores: 88, 86, 84, 79, and 78.

2. Calculate the mean for the following test scores: 100, 81, 79, 78, and 77.

3. Compare what the mean signifies in each of the above examples. In which case does the mean tell you more about the data? Why?

4. The median of a set of data is the value that has an equal number of values above and below it. What is the median score in question 1 above? What is the median in question 2?

5. Is the median a more meaningful statistic in either of the cases listed above? Explain your answer.

10. Sports Averages: Is the Mean a Good Statistic?

Context

sports

Math Topic

mean

Overview

In this activity, students will use what they learned about means in Activity 9. They will now apply that knowledge to sports statistics, the kind of statistics with which many of your students are already familiar.

Objectives

Students will be able to:

- determine which numbers to calculate to find the appropriate mean

- find the significance of a statistic to discover the truly better player

Materials

- one copy of the Activity 10 handout for each student

- calculators

Teaching Notes

Begin your discussion by getting students talking about the use of statistics in sports.

Many students will be familiar with batting averages, points per game, and other such statistics. Be sure that students understand that these per game statistics are means.

Explain that statistics play a particularly important role in team sports because such a range of talent and ability exists on each team. Ask students why these statistics are calculated. Lead them to the conclusion that it can help point out the strengths and weaknesses of a team, and perhaps help to explain why a team won or lost a game. Tell students that in this activity they will complete different activities demonstrating why sports averages are sometimes good, and also why they can sometimes be misleading. Students will then come to their own conclusions about how they feel about sports statistics.

Answers

1. a. Hannah scored 6 points per game.

    b. Phoebe scored 6 points per game.

2. Answers will vary. Propositions that either player is better could be supported by these data since the mean points per game are equal. Students might point out that Hannah is more consistent because Phoebe got a lot of her points in just one great game.

(continued)

10. Sports Averages: Is the Mean a Good Statistic?

3. Here is a completed table:

Player	Games	Points	Points per game	Rebounds	Rebounds per game	Assists	Assists per game
Hannah	18	108	6.0	69	3.8	88	4.9
Phoebe	14	84	6.0	65	4.6	104	7.4

It is clear that Phoebe is contributing more rebounds and assists per game, so the case can be made that Phoebe is the more valuable player.

Extension Activities

- Take students to the gym. Break most students into small teams and have other students be the record keepers. Have team members each shoot a basketball ten times while the record keepers write down the shots and the scores. After everyone is done, return to the classroom and calculate the averages for the individuals as well as the teams.

- Have students use the Internet to find data on a favorite NBA or WNBA player. Have them calculate:

  - the number of games the player played in this season

  - the number of points the player scored in each game

  - the number of rebounds and assists the player made in each game

  - the player's mean points, rebounds, and assists per game this season

Students can then compare their data to decide who they think is the most valuable player this season.

10. Sports Averages: Is the Mean a Good Statistic?

The girls' basketball season has just ended. Two of the players, Hannah and Phoebe, are particularly competitive. They want to know who is the better player. Help them figure it out by answering the questions below.

1. a. Hannah played in 18 games during the season. During this time, she scored 4 points in each of 10 games, 5 points in each of 4 games, 10 points in each of 3 games, and 18 points in 1 game. What is the mean number of points Hannah scored per game?

    b. Phoebe played in 14 games during the season. During this time, she scored 4 points in each of 5 games, 5 points in each of 2 games, 10 points in each of 3 games, 24 points in 1 game, and did not score in 3 games. What is the mean number of points Phoebe scored per game?

2. Based on this information, who do you think is the better player? Explain your answer.

3. Hannah and Phoebe decide that they need to analyze more than just the number of points scored per game in order to tell who is the better player. So, they analyze their rebounds and assists as well. They quickly collect the necessary data and compile it in the following table:

Player	Games	Points	Points per game	Rebounds	Rebounds per game	Assists	Assists per game
Hannah	18	108		69		88	
Phoebe	14	84		65		104	

Complete the table above. Then make a judgment about whether Hannah or Phoebe is the more valuable player, and explain the reasons for your judgment.

11. Baseball: The Sport of Statistics

Context

sports

Math Topic

creating statistics

Overview

Baseball has always had detailed statistics. These statistics have been consistently recorded for about a century. In this activity, students will look at statistics for some great players and try to develop creative statistics to compare their performances.

Objectives

Students will be able to:

- understand and calculate basic baseball statistics

- create new statistics in order to evaluate performance

Materials

- one copy of the Activity 11 handout for each student

- calculators

Teaching Notes

This activity is a good one for inspiring students to think about where statistics come from. In general, they are made up of pieces of raw data. However, if they are to mean anything, they must be constructed purposefully. Baseball is full of purposeful statistics, only a few of which are discussed here.

You will probably want to begin by making sure that everyone understands what all the basic columns in the batting statistics chart mean. Here is an explanation of the abbreviations:

Symbol	Name	Meaning
BA	batting average	equal to hits (H) divided by at bats
AB	at bats	number of times up to bat, less number of bases on balls received, hit by pitch, and other awards of first base or sacrifice hits
R	runs scored	number of times batter safely reaches home plate
H	hits	number of times batter hit safely to reach base
TB	total bases	sum of 1 times 1-base hits, 2 times 2-base hits, 3 times 3-base hits, and 4 times home runs
HR	home runs	hits for 4 bases
RBI	runs batted in	runs driven in through hits or other means
BB	bases on balls	number of times walked to first base by pitcher (4 balls)
SO	strikeouts	number of times struck out by pitcher (3 strikes)
SB	stolen bases	number of base advances prior to batters completing a time at bat

(continued)

11. Baseball: The Sport of Statistics

The Ratio column is explained on the activity sheet.

Let students complete the blank columns when you are sure that they understand the meaning of the columns. Then open up discussion of how to use the basic and created statistics to rate the players. Some of your students, or even your colleagues, will likely be participants in online fantasy baseball leagues where creative statistics are often concocted.

Answers

1. The table is completed below. The batting average gives only the ratio of hits to at bats. It does not measure the quality of those hits—whether they were singles, doubles, triples, or home runs. Nor does it measure a player's ability to advance (steal bases) once on base.

2. The ratio is a measure of the quality of a player's hits (it includes total bases a player hits for), a player's ability to get on base by being walked, and a player's ability to advance by stealing. It also penalizes for strikeouts, since these outs can't help other runners to advance. In short, it is a measure of the player's ability to advance on the bases.

3. Answers will vary. Students may want to experiment with weighting factors on various statistics. You might want to suggest that they come up with creative names for them.

MAJOR LEAGUE BASEBALL NATIONAL LEAGUE BATTING LEADERS FOR 2005 SEASON											
Batters	BA	AB	R	H	TB	HR	RBI	BB	SO	SB	Ratio
D. Lee	0.335	594	120	199	393	46	107	85	109	15	0.566
A. Pujols	0.330	591	129	195	360	41	117	97	65	16	0.593
M. Cabrera	0.323	613	106	198	344	33	116	64	125	1	0.419
T. Helton	0.320	509	92	163	272	20	79	106	80	3	0.489
S. Casey	0.312	529	75	165	224	9	58	48	48	2	0.392
C. Tracy	0.308	503	73	155	278	27	72	35	78	3	0.442
M. Holliday	0.307	479	68	147	242	19	87	36	79	14	0.414
J. Bay	0.306	599	110	183	335	32	101	95	142	21	0.445
B. Clark	0.306	599	94	183	255	13	53	47	55	10	0.398
D. Wright	0.306	575	99	176	301	27	102	72	113	17	0.428

(continued)

11. Baseball: The Sport of Statistics

Extension Activities

- Below is a table containing pitching performance statistics. The column labeled WHIP (walks and hits per innings pitched) is computed as follows:

Ratio = {bases on ball (BB) + hits (H)} / {innings pitched (IP)}

MAJOR LEAGUE BASEBALL NATIONAL LEAGUE PITCHING LEADERS FOR 2005 SEASON										
Pitchers	W	L	Win %	ERA	IP	H	ER	BB	SO	WHIP
R. Clemens	13	8	0.619	1.87	211.1	151	44	62	185	1.008
A. Pettitte	17	9	0.654	2.39	222.1	188	59	41	171	1.030
D. Willis	22	10	0.688	2.63	236.1	213	69	55	170	1.134
P. Martinez	15	8	0.652	2.82	217.0	159	68	47	208	0.949
C. Carpenter	21	5	0.808	2.83	241.2	204	76	51	213	1.055
J. Peavy	13	7	0.650	2.88	203.0	162	65	50	216	0.977
R. Oswalt	20	12	0.625	2.94	241.2	243	79	48	184	1.204
J. Smoltz	14	7	0.667	3.06	229.2	210	78	53	169	1.145
J. Patterson	9	7	0.563	3.13	198.1	172	69	65	185	1.195
C. Zambrano	14	6	0.700	3.26	223.1	170	81	86	202	1.146

Have students study these statistics and decide if Ratio is a good statistic for evaluation of pitching performance. (*Note:* IP—"Innings Pitched"—is given by a number that looks like a decimal, but isn't. 241.1 means 241 innings plus one more out or, actually 241.333 innings. Similarly, 196.2 actually means 196.6.)

- Have groups of students research this season's MLB players on the Internet. Ask each group to collect one player's statistics for all the information shown on the table on the handout, then compile all the data into a table comparing this season's players.

11. Baseball: The Sport of Statistics

You often argue with your friends about who are the best players in Major League Baseball. So you learn about basic baseball statistics, and then use them to make up new statistics to rate hitters for the performance areas you decide are most important. Below is a list of the players you believe were the top offensive players in the National League in 2005.

MAJOR LEAGUE BASEBALL NATIONAL LEAGUE BATTING LEADERS FOR 2005 SEASON											
Batters	BA	AB	R	H	TB	HR	RBI	BB	SO	SB	Ratio
D. Lee		594	120	199	393	46	107	85	109	15	
A. Pujols		591	129	195	360	41	117	97	65	16	
M. Cabrera		613	106	198	344	33	116	64	125	1	
T. Helton		509	92	163	272	20	79	106	80	3	
S. Casey		529	75	165	224	9	58	48	48	2	
C. Tracy		503	73	155	278	27	72	35	78	3	
M. Holliday		479	68	147	242	19	87	36	79	14	
J. Bay		599	110	183	335	32	101	95	142	21	
B. Clark		599	94	183	255	13	53	47	55	10	
D. Wright		575	99	176	301	27	102	72	113	17	

1. Complete the batting average statistic (BA) for each player.

$$BA = \frac{H \text{ (hits)}}{AB \text{ (at bats)}}$$

Why might the batting average be an incomplete measure of a hitter's performance?

2. Complete the ratio statistic for each hitter. This is a made-up statistic using the following equation:

$$\frac{TB + BB + SB - SO}{AB + BB}$$

How might this statistic be a more complete picture of hitting performance than the batting average?

3. Make up your own performance statistic. Explain why you think this will help identify the best players.

12. Comparing Data Statistically

Context

driving

Math Topic

statistics representing two means

Overview

In this activity, students learn about the distribution of data and the term *variability*. They will take two sets of related data, calculate the means of each, and try to draw conclusions based solely on these means. But save these data! In a later study of error in measurements, it will be established that differences in means may not be significant.

Objectives

Students will be able to:

- demonstrate understanding of the difference between a control group and an experimental group in a statistical study

- compare two means and realize that the differences may not be significant

Materials

- one copy of the Activity 12 handout for each student

- calculators

Teaching Notes

Review with students what they have learned in previous activities about means. Ask them if they can think of practical uses for them. Lead students to the discussion of studies, such as when a drug company or the government studies the effectiveness of a new pharmaceutical product.

Explain that in such a study, there are usually two groups. The control group is the group that uses the old product. The experimental group is the group that tries the new product. In such a study, the subjects do not know which group they belong to. Which individual goes into which group is selected randomly to avoid inadvertent skewing of results.

Students may question the need for two groups. Lead them to conclude that such studies are effective. Such a study allows you to gather two sets of data and then find the confidence interval for the mean. In later activities, confidence and significance will be studied in more detail, so have students keep the results of this activity.

Answers

1. control group

2. experimental group

3. Group 1 mean: 6.07

   Group 2 mean: 4.80

(continued)

12. Comparing Data Statistically

4. Based on means alone, Group 1 performed better. However, this is not the end of the story. Emphasize the fact that the variation of the data allows for some probability that the difference in these means is due to chance alone. Save these data for later use.

5. Here lies another major issue in statistics—bias. Any previous training could be expected to bias the results.

Extension Activity

Have students figure out how this type of study may or may not work for a food company taste testing a new product. Have them report on their results.

12. Comparing Data Statistically

Suppose a driver's education teacher has recently returned from a seminar where she learned a new method for teaching students how to parallel park. Before she changes her curriculum, she wants to see if this method really does work better.

She decides to conduct a study. She will randomly choose people from the pool of students signed up for her course and put them into two classes. One class will be taught how to parallel park using the old method. The second group will learn using the new method. At the end of the course, she will rate who parks better and calculate her results.

1. What would the first group be called?

2. What would the second group be called?

Assume that there are 15 students in each class and the teacher rates each on a scale of 1 to 10, with 1 being the lowest score and 10 being the highest. The teacher recorded the scores as follows:

Group 1	Group 2
6, 8, 7, 7, 5, 6, 3, 9, 5, 5, 3, 7, 6, 7, 7	6, 5, 3, 2, 7, 4, 5, 8, 4, 3, 6, 4, 4, 6, 5

3. Find the mean for each group and record your calculations.

4. Which group performed better, based on comparisons of means alone?

5. If three members of the first group and two members of the second group knew how to parallel park before taking the class, would their preexisting ability affect the results?

13. Why What Is Normal Has Deviation

Context

daily life

Math Topic

normal distribution

Overview

The normal distribution is the cherished heart of classical statistics. If a population of series of measurements is distributed normally about a mean, forming the familiar bell-shaped curve, much mathematical magic is possible when performing calculations. Along with the mean, the most useful statistic about normal data is the standard deviation, a measure of the variation or "spread-outness" of the data. The discussion and class examples prepare students to calculate and analyze their data from Activity 2 and Activity 12.

Objectives

Students will be able to:

- identify populations or sets of measurements that are distributed normally

- recognize the bell-shaped curve of the normal distribution

- understand the meaning of standard deviation

Materials

- one copy of the Activity 13 handout for each student

- scientific calculators, graphing calculators, or computers

- Reaction-time data from Activity 2; if students have not completed Activity 2, give them the sample data set from the Teacher's Page, Activity 8.

Teaching Notes

Review with students what they have learned in previous activities regarding mathematical means. Explain that the term *standard deviation* refers to what underlies that mean, allowing people to better understand the true meaning of the data.

A good way to help students envision a normal distribution is to have them imagine data grouped evenly around a mean in the middle, with many of the data points near the mean. Data sets that come from measurements that are unbiased and not skewed in some manner are good candidates for normal distribution. Some examples include height of adults, number of hairs on high-school students' heads, and even the reaction-time measurements from Activity 2. But explain that most sets of data are not normal.

(continued)

13. Why What Is Normal Has Deviation

Standard deviation is used to explain how clustered the data are around the mean. When the data are grouped tightly around the mean, like the sample reaction-time measurements, the standard deviation is low. When the data are more spread out, the standard deviation is large. The standard deviation is a statistic that allows you to interpret the spread of data and calculate the uncertainty, or error, in the measurements.

Students should use calculators to complete question 3. (If they have not completed Activity 2, give students the sample data set from the Teacher's Page in Activity 8.) When students have completed the question, ask the whole class to compare the results and discuss their meaning.

Answers

1. The curve is bell-shaped with 50 of 100 successes in the middle and a maximum value of 8 of 100 occurrences. The curve has tails that fall quickly to 0 occurrences above 60 successes and below 40 successes.

2. Answers will vary, but should be about 50 successes, which is the mean number of successes.

3. Answers will vary. Measurements that involve random fluctuation about a mean will produce normal distributions. See examples in the Teaching Notes.

4. For the sample set from Activity 8, $s = 0.02$. In general terms, this means that about two-thirds of the measurements will fall within 0.02 seconds above or below the mean. Answers will vary.

Extension Activities

Have students produce a series of graphs similar to the one on the activity sheet. Have them vary the number of coins tossed, number of repetitions, and probability of successes.

13. Why What Is Normal Has Deviation

You and a few friends are bored. Things are so bad, you've been sitting around tossing coins and guessing whether heads or tails will come up. The results seem pretty random. Then you say, "I wonder what would happen if you tossed 100 coins at the same time? How many of them would come up heads, and how many would come up tails?"

So you get two rolls of pennies, for a total of 100 coins, and clear a big space to toss the coins. You're all ready to start tossing coins when someone says, "Hey, how are we going to keep track of the heads and tails? We'll never remember how many come up heads each time unless we write it down." So now you need a record-keeping form, too. You decide to call each coin that lands with heads up a success, and you record the number of successes each time you toss the 100 coins. You end up tossing the 100 coins 100 times. You then graph the results. Here is the graph you end up with.

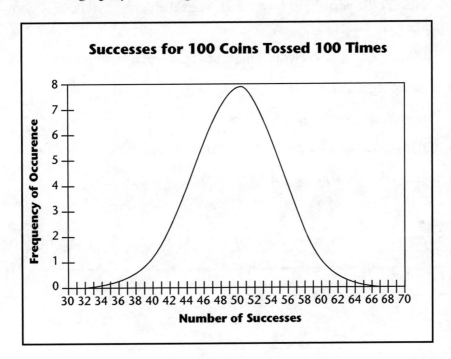

Successes for 100 Coins Tossed 100 Times

Frequency of Occurence (y-axis: 0 to 8)

Number of Successes (x-axis: 30 32 34 36 38 40 42 44 46 48 50 52 54 56 58 60 62 64 66 68 70)

1. The graph above shows a normal distribution. The *y*-axis shows how frequently the number of successes on the *x*-axis occurs if getting heads is a success. Describe its shape. What number of successes was most common? How often did it occur? (This is the maximum value of the distribution.)

(continued)

13. Why What Is Normal Has Deviation

2.   What would you guess was the mean number of successes?

3.   Think about data sets in real life that follow a normal distribution. List as many as you can.

   Standard deviation tells you about how much your data are spread out. Follow these steps to calculate standard deviation.

   Step 1: Subtract the mean of your whole set from each individual value in the set.

   Step 2: Square these differences.

   Step 3: Sum up all those squared values.

   Step 4: Divide that sum by one less than the total number of data values.

   Step 5: Then take the square root of the resulting value.

4.   Copy the data you compiled in Activity 2 here. What is the standard deviation?

   $s =$ _____

14. Dates on Coins

Context

money

Math Topic

frequency distribution

Overview

In this activity, you or the students will gather as many coins as you can and determine the frequency distribution of dates on the coins. Students will use this information to draw a graph.

Objectives

Students will be able to:

- graph data

- write about their discoveries

Materials

- one copy of the Activity 14 handout for each student

- calculators

Teaching Notes

To get students discussing the concept of frequencies, begin by talking about its purpose. Students should be able to determine that large sets of data can be reported by frequencies rather than individual observations. For example, reporting to the school how many students would like pizza for lunch would be easier than providing it with the list of students who selected pizza as their top lunch choice. Frequencies can also be reported in percentages: "Eighty percent of students in this school would like pizza for lunch."

In this activity, students will discover the frequency of dates on coins and compile their data into a graph. When the information is displayed in this manner, students will be able to see which dates were more common on coins and which were less common. Students should then be able to discuss their findings and come up with reasons for the variation in dates. You can choose to bring in coins for the class to share, tell students to bring in coins on a specific day, or have students work on this activity at home using their own coins.

Answers

Answers will vary according to the individual results of the students.

Extension Activity

Have students discuss a word game such as Scrabble®. Those who have played the game will probably recall the different point values of each letter. They may also recall that the less common letters, such as Q and W, have higher point values than letters such as A and S. Allow students to conclude that because the less common letters appear in words with less frequency, they are more difficult to use and are therefore worth more points. Have students write about their discoveries.

14. Dates on Coins

Collect as many coins dated from 1998 through 2007 as you can and place them in a bag or pile. Look at the coins one at a time, checking the date as you look at them. Sort the coins into piles for each year. Then count each pile and enter the total number of coins in the Frequency column. When you find a date not on the chart below, set the coin aside.

Year	Frequency	Year	Frequency
1998		2003	
1999		2004	
2000		2005	
2001		2006	
2002		2007	

The lines you draw on the graph will represent the number of coins in each pile. For example, if you had nine coins from 2002, that line would look like this:

2002 ────────────
      2   4   6   8   10   12   14   16

1. Use the chart below to draw your graph.

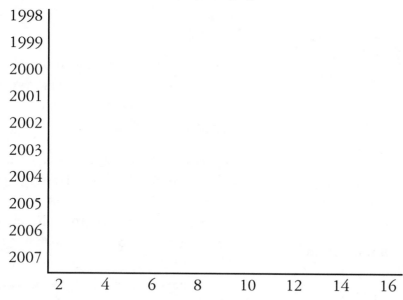

2. Describe the shape and other important features of this graph.

15. Letters of Fortune

Context

television

Math Topic

relative frequency

Overview

In this activity, students will analyze a sample of English words to see how often each letter tends to appear.

Objectives

Students will be able to:

- analyze the relative frequency of values in a sample

Materials

- one copy of the Activity 15 handout for each student

- calculators

Teaching Notes

Start by asking students if they ever watch *Wheel of Fortune,* or play games such as Hangman or Scrabble. Ask whether they have ever noticed any patterns in the frequency with which letters appear. In Scrabble, are there the same number of tiles for each letter? (No; for some letters there are eight or nine tiles; for some, only one or two.) Ask students to speculate on why this might be.

Have students work in pairs or individually to tally the letters in the first three paragraphs on the handout. If students work in pairs, direct each student to complete the letter distribution table. When both students have completed the tally, they can compare totals to check the accuracy of their counts. If the two tallies do not agree, then both students should tally the letters again, until both tallies are the same.

(continued)

15. Letters of Fortune

Answers

Letter Frequency Chart

Letter	Total	Relative Frequency	Letter	Total	Relative Frequency
A	31	0.068	N	30	0.066
B	1	0.002	O	52	0.114
C	16	0.035	P	6	0.013
D	11	0.024	Q	0	0
E	53	0.116	R	31	0.068
F	12	0.026	S	36	0.079
G	9	0.020	T	42	0.092
H	29	0.063	U	22	0.048
I	16	0.035	V	2	0.004
J	2	0.004	W	15	0.033
K	4	0.009	X	0	0
L	12	0.026	Y	16	0.035
M	9	0.020	Z	0	0
				457	

1. The sample included 457 letters.

2. Answers will vary. Students should indicate that the sample was reasonably large. Some may comment that the fact that it was addressed to "you" may have skewed the results, as the letter *Y* appeared frequently in the sample—perhaps more frequently than in most English words.

3. The most frequent letters, in order of frequency, are *E, O, T, S, R, A, N, H, U, C, I,* and *Y*.

4. The least frequent letters in this sample are *Q, X, Z* (0 occurrences); *B* (1 occurrence); and *J, V* (2 occurrences each).

5. The first choice vowel should be *E*, with a relative frequency of 0.116.

6. The first choice consonant should be *T*, with a relative frequency of 0.092.

7. The statement that the top ten letters account for about 75% of the letters used in writing English is true.

Extension Activities

- Have students create a scatter plot to show the data on the table. How does a scatter plot help show the results of the analysis?

- Based on their findings in the activity, have students modify the rules of hangman by assigning different values to different letters according to the relative frequency with which they are used. Then play Hangman using the new rules to see how well this variation works.

15. Letters of Fortune

You just got a call from the producers of the show *Wheel of Fortune.* You're going to be a contestant on the show! A friend suggests that you do some research first to improve your chances of winning.

"Research?" you say. "How can I do research? I won't know what they're going to ask, so how can research help?"

"You can research letters," your friend says. "I've noticed that some letters show up more often than others—just like when you play Hangman. And some don't show up often at all. If you find out which letters are used most, you can make sure you don't waste your letters on the show."

You realize that this makes sense, so you sit down together and come up with a plan. You will take the first three paragraphs of this activity and see how often each letter is used. Then you will calculate the relative frequency with which each letter appears and develop a "top ten" list of letters to use on *Wheel of Fortune.*

Start at the beginning of the first paragraph, with the words "You just got a call. . . ." Make a tally mark in the appropriate space in the table on the next page each time a letter is used. Continue tallying until the end of the third paragraph, with the words ". . . you don't waste letters on the show." When you have tallied all the letters in all three paragraphs, total the number of times each letter appeared. Write each total in the appropriate space in the table. Then add up all the totals to find the total number of letters analyzed.

Now calculate the relative frequency of each letter. To do this, divide the number of times each letter appeared by the total number of letters in the sample. For example, if you had sampled 150 letters and the letter *E* showed up 17 times, you would find $17 \div 150$, for a relative frequency of 0.113.

Once you have completed the table, use it to answer the questions on the next page.

(continued)

15. Letters of Fortune

Letter Frequency Chart

Letter	Tally	Total	Relative Frequency
A			
B			
C			
D			
E			
F			
G			
H			
I			
J			
K			
L			
M			
N			
O			
P			
Q			
R			
S			
T			
U			
V			
W			
X			
Y			
Z			
	Total letters in sample:		

(continued)

15. Letters of Fortune

1. How large a sample of letters did you examine?

2. Was that sample large enough to be able to predict how often the letters appear in most English words? Explain your answer.

3. Based on your sample, which ten letters appear most often in English words? List them in order of frequency.

4. Based on your sample, which five letters appear least often in English words?

5. Which vowel should be your first choice when you play *Wheel of Fortune*?

6. Which consonant should be your first choice?

7. Based on your findings, is the statement below true or false?

    "The top ten letters account for about 75% of the letters used in writing English."

16. Class Statistics Quiz

Context

education

Math Topic

frequency distribution, mean

Overview

In this two-part activity, students will complete an eight-question multiple-choice quiz based on statistics concepts they have learned up to this point. They will then analyze their responses and compile the statistics on their statistics quiz.

Objectives

Students will be able to:

- check knowledge on basic statistics concepts

- use quiz data to determine the frequency distribution and mean of class scores

Materials

- copies of the three Activity 16 handouts for each student

- calculators

Teaching Notes

You can choose to review statistics material with your students prior to the quiz. It is important that you cover all the basic concepts included in the eight quiz questions. The quiz is not designed to be difficult; it is provided to give students a chance to analyze data from their own class performance.

Distribute the two-page quiz first and have students complete it. Afterward, go over the quiz very thoroughly. The answers on the next page include extra detail about each question. Make sure all students understand the correct response for any items they missed.

Prior to beginning the second part of the activity, you should have a discussion on frequency distribution. Begin by explaining that this is often a good choice for presenting data. Class results will show the number of students scoring 0, 1, 2, 3, 4, 5, 6, 7, and 8 correct answers. Have students construct histograms on the distribution of quiz results, calculate the mean, and decide the scores that deserve grades of A, B, or C.

Make sure students know how to calculate the mean given the frequency distribution. They simply need to add the products of multiplying each score times the frequency with which it occurred, and then divide by the total number of quizzes. Calculators will be helpful here.

(continued)

16. Class Statistics Quiz

Answers

Quiz

1. The correct answer is *c*. Sometimes a person who suffers an unfortunate mishap is referred to as a "statistic." But this is not the mathematical usage so choice *a* is incorrect.

2. The correct answer is *b*. Neither mean nor standard deviation is a landmark, since these must be calculated. So neither *a* nor *c* is correct.

3. The correct answer is *a*. Average and mean are usually equivalent, but sometimes average refers to something that is not technically a mean, such as baseball batting average.

4. The correct answer is *c*.

5. The correct answer is *c*.

6. The correct answer is *b*.

7. The correct answer is *b*. The 6% margin of error can swing either way about the reported result. Therefore, it is not clear at a 95% confidence level which candidate will win.

8. The correct answer is *a*. This is the trickiest question. It must be read carefully. The probability that the 95% confidence interval does not contain the actual mean is only 5%. In other words, the confidence interval calculated from the sample has a 5% or lower chance of being wrong for the whole population.

Analysis

1. Distribution will depend on student results.

2. Histogram will depend on student results.

3. Mean will depend on student results. Make sure that students know how to use the frequency distribution to calculate the mean.

4. Leave this subjective judgment up to students. They might require a score 1, 2, or 3 points above the mean to receive an A. On the other hand, if everyone scored 6 or above, they may rightly feel everyone should get an A. But if no one makes 6, then perhaps no one should get an A. Let the class tell you how they feel about their scores.

Extension Activity

Have students make up their own statistics quizzes and analyze the results.

16. Class Statistics Quiz

Circle the letter of the response that best answers the question or completes the sentence.

1. A statistic is

   a. a person who suffers a serious injury.

   b. a set of data.

   c. a number that gives a measurement of a set of data.

   d. all of the above

2. The following are data landmarks:

   a. mean, median, and mode.

   b. maximum, minimum, range, median, and mode.

   c. average, mean, and standard deviation.

   d. all of the above

3. Average and mean

   a. are usually equivalent, but average sometimes refers to quantities that are not really means.

   b. are always exactly the same thing.

   c. are defined as the middle values.

   d. cannot be calculated at the same time.

4. Over five evenings, a student watches television for 1.0 hour, 2.5 hours, 3.0 hours, 2.0 hours, and 1.5 hours. The mean number of hours the student watches television per evening is

   a. 1.0 hour.

   b. 1.5 hours.

   c. 2.0 hours.

   d. 2.5 hours.

(continued)

16. Class Statistics Quiz

5. Standard deviation is

    a. a statistic measuring the "spread-outness" of a set of data.

    b. often used to calculate errors in sets of measurements and confidence limits.

    c. both *a* and *b*

    d. neither *a* nor *b*

6. A normal distribution

    a. cannot have outliers.

    b. arises from randomly distributed measurements and is represented by a bell-shaped curve with the mean of the distribution at the center and data spread evenly about the mean.

    c. arises from biased measurements and is represented by a bell-shaped curve with data skewed to one side of the mean.

    d. none of the above

7. A political poll says that candidate *y* is leading candidate *z* 54% to 46% with a margin of error of 6%. Then you are 95% sure that

    a. candidate *y* is favored by the total voting population by 54% to 60% and candidate *z* is favored by 40% to 46%.

    b. candidate *y* is favored by the total voting population by 48% to 60% and candidate *z* is favored by 40% to 52%.

    c. candidate *y* will win.

    d. candidate *z* will win.

8. When you have a sample of data from a very large population, a confidence interval of 95% tells you

    a. the range of values around the sample mean within which the probability that actual mean of the population is not included is less than 5%.

    b. the range of values around the sample mean within which the probability that actual mean of the population is not included is less than 95%.

    c. the amount of time you are confident that your sample contains 95% of the data from the actual population.

    d. none of the above

(continued)

16. Class Statistics Quiz

Analysis

The class quiz on the basics of statistics is over. Now you have a chance to use class results from the quiz to learn more about statistics.

1. Complete the table below showing the frequency distribution for your class statistics quiz.

Number of questions correct	Number of students scoring this number
0	
1	
2	
3	
4	
5	
6	
7	
8	

2. Make a histogram of these data below.

Number of Occurrences vs. Number Correct

```
20

15

10

 5

 0
     0   1   2   3   4   5   6   7   8
```

3. What is the mean of the class quiz scores?

4. Did your class do well on the quiz? What score should be needed for an A?

17. Yao Ming: How Many Points per Game?

Context

sports

Math Topic

mean and confidence interval

Overview

This activity is a recreational departure in the application of confidence intervals. Students will calculate the mean for a sample of Yao Ming's scoring statistics during the 2004–05 season with the Houston Rockets. They will then calculate the 95% confidence interval in an attempt to predict the likely mean number of points Yao Ming will score in any given game.

Objectives

Students will be able to:

- calculate a mean and 95% confidence interval

- apply statistics to predict sports performances

Materials

- one copy of the Activity 17 handout for each student

- calculators or computers with spreadsheet software

Teaching Notes

Confidence intervals are revisited in this activity. Students may ask why 95% has been chosen as the limit. Explain that this is how statistics and probability are linked— you can never be absolutely sure that a small sample will return statistics exactly equal to those for an entire population. But you are able to choose reasonable limits within which there is a high probability that your sample statistics accurately reflect the whole. A 95% confidence interval is the range within which the probability that the actual population mean is outside this range is less than 5%. While other stricter, or looser, limits may be chosen, the 95% interval is very often preferred by statisticians.

There are two ways to approach the Yao Ming scoring data. One approach is to focus on per game statistics; the other is to focus on points per minute played. The per game statistics measure Yao Ming's output for total games, while the per minute statistics measure his output only while playing. Discuss possible differences in the two approaches with the students. You may have all of the students do the calculations both ways or split up the class to do one or the other.

It is highly desirable for all the calculations to be performed with calculators that have statistical functions or computers with statistics-capable software. You should instruct students in proper use of the technology. Calculators with statistical functions will require only that students punch in the data. The mean and standard deviation will then be available at the touch of a button. Otherwise, your students will get bogged down in the fairly tedious calculations.

(continued)

17. Yao Ming: How Many Points per Game?

Answers

1.

Opposing team	Date of game	Points scored	Minutes played	Points per minute
Phoenix Suns	March 11	27	35	0.77
Sacramento Kings	March 13	17	30	0.57
Golden State Warriors	March 14	14	29	0.48
Portland Trail Blazers	March 16	13	21	0.62
Boston Celtics	March 18	18	28	0.64
Minnesota Timberwolves	March 20	21	31	0.68
Miami Heat	March 22	12	25	0.48
Cleveland Cavaliers	March 24	13	24	0.54
New Orleans/ Oklahoma City Hornets	March 25	12	24	0.50
San Antonio Spurs	March 27	18	31	0.58
Utah Jazz	March 28	15	29	0.52
Phoenix Suns	April 3	19	33	0.58
Golden State Warriors	April 5	23	38	0.61
Los Angeles Lakers	April 7	21	32	0.66
Phoenix Suns	April 9	10	25	0.40
Seattle SuperSonics	April 11	20	31	0.65
Memphis Grizzlies	April 13	28	33	0.85
Denver Nuggets	April 16	13	23	0.57
Los Angeles Clippers	April 18	24	28	0.86
Seattle SuperSonics	April 20	13	23	0.57
Mean		17.6	28.7	0.60
Standard deviation		5.2	4.5	0.12
Confidence interval		2.3	2.0	0.05

(continued)

17. Yao Ming: How Many Points per Game?

2. Students may notice that the confidence interval in Yao Ming's point production is small—about 18 ± 2 points. Based on this sample of games, there is a 95% probability that the mean number of points Yao Ming scores in a game is between 16 and 20. The confidence interval for his per minute played point production is about 0.60 ± 0.05 points.

Extension Activity

Have students use the Internet to look up statistics for Michael Jordan, who was a leading scorer for the National Basketball Association (NBA). Have them perform the same calculations they did with the Yao Ming data, compare the results with those data, and report their findings.

17. Yao Ming: How Many Points per Game?

Yao Ming of the NBA's Houston Rockets scores consistent per game point totals. Investigate his performance by calculating the mean, standard deviation, and 95% confidence interval for a sample of 20 regular season games.

1. Complete the table below according to your teacher's instructions. Calculate the mean, standard deviation, and confidence interval for the columns you are assigned.

Opposing team	Date of game	Points scored	Minutes played	Points per minute
Phoenix Suns	March 11	27	35	
Sacramento Kings	March 13	17	30	
Golden State Warriors	March 14	14	29	
Portland Trail Blazers	March 16	13	21	
Boston Celtics	March 18	18	28	
Minnesota Timberwolves	March 20	21	31	
Miami Heat	March 22	12	25	
Cleveland Cavaliers	March 24	13	24	
New Orleans/ Oklahoma City Hornets	March 25	12	24	
San Antonio Spurs	March 27	18	31	
Utah Jazz	March 28	15	29	
Phoenix Suns	April 3	19	33	
Golden State Warriors	April 5	23	38	
Los Angeles Lakers	April 7	21	32	
Phoenix Suns	April 9	10	25	
Seattle SuperSonics	April 11	20	31	
Memphis Grizzlies	April 13	28	33	
Denver Nuggets	April 16	13	23	
Los Angeles Clippers	April 18	24	28	
Seattle SuperSonics	April 20	13	23	
Mean				
Standard deviation				
Confidence interval				

2. Interpret your results. How do the points per game statistics differ from the points per minute statistics?

18. Growth of the Internet

Context

computers

Math Topic

exponential growth

Overview

Statistics are very important for discovering, measuring, and predicting real-world trends. And the most interesting trends are not linear. Trends that involve growth can usually be expressed by such statements as "The population is growing by 18% per year." Such growth is called *geometric* or *exponential* because it is based on a larger population each year. In this activity, students will look at a graph illustrating the phenomenal geometric growth of the Internet. They will attempt to determine its recent growth rate and then predict its future growth.

Objectives

Students will be able to:

- interpret graphed data in order to identify growth trends

- calculate growth rate and predict future growth from graphed data through qualitative regression analysis

Materials

- copies of the two Activity 18 handouts for each student

Teaching Notes

The Internet provides a classic model for geometric growth. In this activity, students will examine real data from The Internet Domain Survey. The results of these counts of Internet host sites have been published since 1987 by Network Wizards. The web site from which the data were obtained (www.isc.org) includes documents explaining the methods used to count host sites. The methods themselves include interesting statistical procedures. As an extension, have students examine this site.

Without dwelling on the mathematical details, students will simply draw the best smooth curve through the data that they can, and then attempt to discover the growth rate. Explain to them that this is the process of nonlinear regression—finding the best curve that "fits" a set of nonlinear experimental data.

Answers

1. On the next page is a graph of the Internet host site data with a computer-derived regression line imposed. Students should be given credit if they come up with anything close that does not have a jerky appearance. The idea is that they make a smooth curve that comes as close as possible to all of the data points.

   The mathematical details are far too cumbersome for students to work out

(continued)

18. Growth of the Internet

by hand at this level. But, if students have access to graphing calculators and computers, they can collect the raw data from the web site and do the analysis themselves. The computer-derived curve is approximated by the function $y = 37,300e^{0.69x}$. The correlation coefficient, $R^2 = 0.99$, is very close to 1.00, indicating that the correlation is excellent. Students need not be bothered with any of these details. However, you should notice that $e^{0.69x}$ represents a continuous annualized growth rate of 69%, leading to a doubling each year.

Internet Host Sites vs. Years Since July 1994

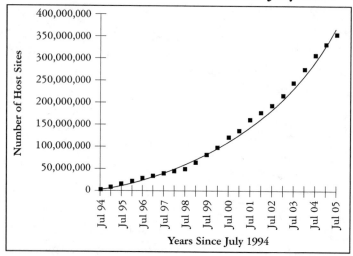

2. The graph is low and flat on the left side, but rises faster and faster until it is very steep on the right side as the time since July 1994 approaches 10 years.

3. The correct answer is 100%. However, you might want to discuss continuous growth rate at this point. Because of growth on the amount a quantity has grown (compound growth), a continuous growth rate of 69% will give you annual doubling. In finance, this is known as compound interest, an extremely important concept.

4. Students must figure out a way to see a pattern in growth for constant increments of time. The beauty of this exponential curve is that any two points that are 1 year apart represents a doubling of the number of sites! The doubling time is about one year.

5. 150,000,000

Extension Activity

Have students visit the web site www.isc.org and report on the methods used to determine the size of the Internet. Interesting statistical sampling techniques are used to verify the domain name count. Students should be able to describe these procedures.

18. Growth of the Internet

Your company wants to expand operations on the Internet, and you are assigned to report on its growth. You find that a good way to chart the growth of the Internet is to study the number of host sites over time. The data you find are shown in the graph below.

1. On the graph below, sketch the best smooth curve possible connecting the points.

Internet Host Sites vs. Years Since July 1994

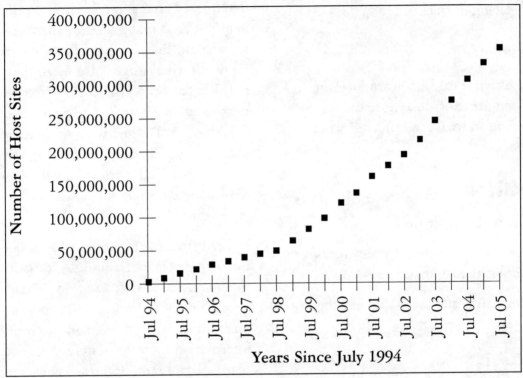

2. What are the important features of the curve you have sketched?

3. In order to double, by what percentage must a quantity increase?

4. At the top of the curve, how many years do your data show that it takes the number of host sites on the Internet to double?

5. Estimate the number of Internet host sites on 01/01/01 from this data.

Real-Life Math: Data Analysis

19. Misuse of Statistics: Advertising Claims

Context

media

Math Topic

interpretation of statistics

Overview

In this activity, students learn how the media and other organizations can manipulate statistics to influence what we believe.

Objectives

Students will be able to:

- see through different presentations to gather the right statistics

- determine statistical misuses and argue against them

Materials

- one copy of the Activity 19 handout for each student

Teaching Notes

Ask students to discuss some medical commercials they have seen on television. These commercials may include phrases like "four out of five dentists agree," "nothing has been proven more effective," and "more hospitals choose."

Ask students what these phrases mean to them. While some students may know that "four out of five dentists agree" can be said after asking only five dentists, most students may believe that this is a statistic determined after interviewing countless dentists. And many students may not understand that "nothing has been proven more effective" might also mean that nothing has been proven less effective either, making all the products advertised equal in effectiveness. Discuss with students that these phrases are developed by advertising agents whose job is to sell the product. While these phrases are not lies, they can be misleading.

Lead students in a discussion about the importance of understanding statistics. They should conclude that a person with a good knowledge of statistics can see beyond the misleading catchphrases and interpret the information more accurately. This makes that person a more informed consumer.

Finally, discuss with students that medical commercials are not the only things that need to be scrutinized. Discuss how political polling results, among other things, can also be manipulated.

(continued)

19. Misuse of Statistics: Advertising Claims

Answers

1. Answers will vary but may include "Nothing works better than Pearly White."

2. Answers will vary but may include "Nine out of ten people prefer Pearly White."

3. Answers will vary but may include "Twice as many people chose Pearly White last month alone."

4. Answers will vary but may include "More hospitals choose Pearly White than any other whitening toothpaste."

5. Answers will vary but may include "More dentists recommend Pearly White to their patients."

Extension Activity

Have students design a commercial for Pearly White using any of the catchphrases they came up with in the activity.

19. Misuse of Statistics: Advertising Claims

You have just been placed in charge of advertising for a new kind of whitening toothpaste, Pearly White. An assigned group of workers has compiled the following statistics about Pearly White. You can manipulate these statistics in any manner you like to sell the product. Come up with a catchphrase for each of the following statistics.

1. Several studies were conducted to test the effectiveness of Pearly White compared to other whitening toothpastes. The studies showed that Pearly White worked about as well as all other whitening toothpastes tested.

2. Surveyors asked ten people to try Pearly White, and nine of them said it was the best whitening toothpaste they had ever tried. However, the study did not ask them if they had ever tried any whitening toothpastes before.

3. Sales of Pearly White have doubled in the past month because the Pearly White Corporation mailed out coupons for a dollar off the product.

4. A majority of hospitals use Pearly White because the Pearly White Corporation provides its toothpaste to hospitals at no charge.

5. More dentists give Pearly White to their patients because the Pearly White Corporation provides free samples to dentists.

20. Misuse of Statistics: Goofy Graphs

Context

media

Math Topic

interpretation of statistics

Overview

In this activity, students learn how graphs that appear in newspapers and other media can be designed to be misleading.

Objectives

Students will be able to:

- read the true meanings of graphed statistics

- determine misrepresentations in graphs and argue against them

Materials

- copies of the three Activity 20 handouts for each student

Teaching Notes

Have students discuss graphs they may have seen online or in newspapers recently. These may include graphs related to political polling, crime rates, or national budget matters.

Ask students how carefully they look at these graphs. Are they likely simply to look at the graph itself, or do they take the time to read the information it presents in order to come to their own conclusions?

Continue the discussion from Activity 19 on the importance of understanding statistics. Students should conclude that the more carefully they analyze a graph, the less likely they are to be persuaded by a flashy presentation.

Answers

1. Students should conclude that the graph is not fair because it begins at 20, making Michael's lead appear bigger.

2. a. The other 30 must have been undecided.

   b. no

3. It would make the graph less valuable, since Michael supports an issue important to football players.

4. The new graph should make the race appear more even, since Michael's lead will not be as large and many voters will be shown as undecided.

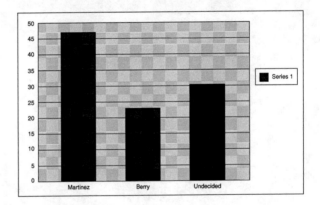

(continued)

 Real-Life Math: Data Analysis

20. Misuse of Statistics: Goofy Graphs

5. Answers will vary, but they may include writing another article for the school newspaper using their new graph, making the newspaper print a correction, or having Keisha respond to the article in a speech.

Extension Activity

Have students design a graph to persuade the school board to add watermelon-seed spitting as a sport for the school. They can manipulate the following statistics in a manner they see fit. Out of 50 students surveyed:

- 17 students would like watermelon-seed spitting to become a school sport

- 10 students said it was a stupid idea

- 23 students didn't care

20. Misuse of Statistics: Goofy Graphs

Keisha Berry is running for student council president against Michael Martinez. Michael's campaign manager has just published an article with a graph in the school newspaper. The information in the article and the graph are hurting Keisha's campaign. Read the information from the school newspaper and then answer the questions that follow. Use this information to determine the best course of action for Keisha.

More Students Prefer Martinez

A poll taken after lunch yesterday showed that if the election were held now, more students would vote for Michael Martinez. A survey of 100 students revealed that 47 would vote for Michael, while only 23 said they would vote for his opponent, Keisha Berry.

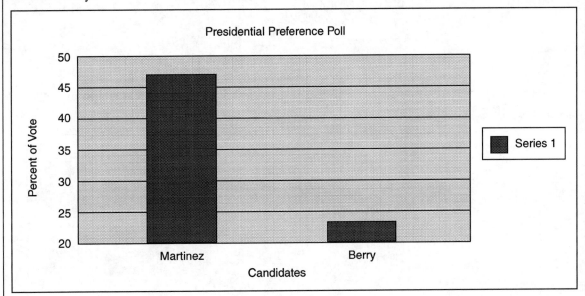

An interview with Martinez revealed that he was happy with the results, crediting his stand on more money for the football team for his lead. Berry could not be reached for comment before the paper was printed.

---

1. Does this graph look fair? Why or why not?

(continued)

20. Misuse of Statistics: Goofy Graphs

2. a. If 100 students were polled and only 70 picked a favorite candidate, what happened to the other 30?

    b. Is it fair that those 30 students were not included in the graph?

3. If you knew that half the students polled were football players, would that be important to the value of the graph?

4. Draw the graph again, this time including undecided voters and starting at 0 instead of 20.

    Describe your new results.

5. What should Keisha do to respond to this article?

Share Your Bright Ideas

We want to hear from you!

Your name_____Date_____

School name_____

School address_____

City _____State _____Zip_____Phone number (_____)_____

Grade level(s) taught_____Subject area(s) taught_____

Where did you purchase this publication?_____

In what month do you purchase a majority of your supplements?_____

What moneys were used to purchase this product?

____School supplemental budget   ____Federal/state funding   ____Personal

Please "grade" this Walch publication in the following areas:

Quality of service you received when purchasing .... A	B	C	D
Ease of use... A	B	C	D
Quality of content... A	B	C	D
Page layout .... A	B	C	D
Organization of material .... A	B	C	D
Suitability for grade level .... A	B	C	D
Instructional value... A	B	C	D

COMMENTS:_____

_____

What specific supplemental materials would help you meet your current—or future—instructional needs?

_____

Have you used other Walch publications? If so, which ones?_____

May we use your comments in upcoming communications?   ____Yes   ____No

Please **FAX** this completed form to **888-991-5755**, or mail it to

Customer Service, Walch Publishing, P. O. Box 658, Portland, ME 04104-0658

We will send you a **FREE GIFT** in appreciation of your feedback. **THANK YOU!**